Learn to Video Call with Kids

Learn to Video Call with Kids

A Guide to Skype and FaceTime to Video Call with Children

Lillian Tibbles, PhD

Platinum Coast Press
Naples, Florida

Learn to Video Call with Kids

Published by
Platinum Coast Press
6017 Pine Ridge Road, Ste. 156
Naples, FL 34119-3956

Cover and interior design by Pat Shapiro,
Flying Colors of Naples, Inc.

ISBN: 978-0-9898818-0-7
LCCN 2013954628
Printed in the United States of America

For Ken, my husband and best friend;
my daughter Lynn and her husband Tim;
my son Chris and his wife Joelle;
and all my wonderful grandchildren.

Our world is changing. Families are separated by thousands of miles.
Video call is a powerful communication tool that is used worldwide
by millions of people. Skype, FaceTime, and other programs
used to video call are free, easy to use, and fun.
This book will help you get connected and stay safe online.

Contents

Conclusion

Author's Note

One of the most exciting days of my life was the day that I learned I would become a grandmother. That excitement was quickly followed by a nagging sense of guilt. Unlike my own mother, who lived just a short distance from me when my children were young, I would be more than three thousand miles from my daughter and new grandson. Frequent travel of that distance would be unlikely. That is when I discovered Skype and video call.

My grandson's first smiles and first attempts at sitting, crawling, and walking were all enjoyed during video calls. It wasn't long before our grandson recognized our voices, and it was even more exciting when he recognized our faces on the computer screen. The first time our grandson spotted us on the computer monitor and crawled across the room to see us, we were thrilled. Soon he was running to see us on the screen, blowing kisses, and waving hello and good-bye.

Fast-forward two years to an airport terminal. Our grandson is running toward my husband and me shouting, "Grandma, Grandpa!" before jumping into our waiting arms. When we are able to be together in person, our grandchildren are far more comfortable with us than would have been possible if they had not been interacting with us on Skype. The emotional connection that we developed with video calls has been priceless. This experience helped me to realize that a simple guide to video call with children would be of value to parents, grandparents, and all others who cannot be with the children they love.

Three generations of my family have participated in the development of this book: my husband, Ken, my daughter, Lynn, my son, Chris, their spouses, and my grandchildren. The joy of being able to see their faces and chat with them across many miles, was the reason this book was written. Many of the ideas in this book came from those video calls, as we all experimented with this wonderful technology. I thank them for all those precious times that they shared with me.

Finally, many thanks to my copy editor Holly Monteith, and to Pat Shapiro, my designer and longtime friend, who used her talents to work with me to develop the cover and interior design for this book. I would also like to thank Bev Kirk for her graphic assistance with the January 2015 Version. I am very grateful to them all.

Introduction

What You Will Find in This Book

If you are separated from family or friends, then this book is for you. It will help you learn to video call with success. It will introduce you to the equipment you need, what children will find fun to do at different ages, what toys and games are fun to use, and how you can celebrate special occasions and holidays together with a video call. It's easy, fun, and far better than communicating by telephone.

I will help you to see how simple the process of making a video call can be and this should spark some new ideas about how to enjoy this wonderful technology with children. This book is not just for parents and grandparents but for all who want to enjoy seeing as well as hearing the voice of a child whom they love.

This book will help you to:

- learn about equipment needed to make a video call
- download and set up Skype's video call application
- learn about FaceTime to make video calls
- communicate face-to-face via the Internet
- stay safe and keep children safe while using video call
- learn about specific developmental milestones in young children
- prepare for video calls with children of different ages
- explore simple, fun activities to share while calling children
- celebrate birthdays, holidays, and other special occasions with video call

A video call from one household to another is free.

Included in this book are the common technical terms that you should know and pictures of the equipment that you will need. It won't be long before you will feel comfortable with the language and equipment of video calling.

Video call is a powerful communication tool that is used by millions of people worldwide. Applications or programs such as Skype and FaceTime are fun and easy to use.

Part I
Video Call

Video call is much more fun than a telephone call—it is simply talking face-to-face using a computer or mobile device. With a video call, ***you can see the person you are calling as well as hear his voice.*** Once you have video called a few times, you will feel as comfortable with it as you would sitting in the same room and chatting with the person. Video call is sometimes called video chat.

VIDEO CALLING MAKES THE DISTANCE BETWEEN FAMILIES SEEM SO MUCH SHORTER—IT CAN DO THAT FOR YOU.

With children, you get to enjoy their every move, their expressions, and their rapid changes as they grow. Whole families who live in different countries or different locations within a country can connect to celebrate birthdays and holidays or just to be together. ***Video calls can be made indoors or outdoors.*** You can take computers and mobile devices outdoors to share the joy of building a snowman, playing in the sand, or riding a bicycle.

Many hospitals in Europe, in Spain, Italy, Poland, and Germany, have installed video call devices in maternity wards to give new mothers the opportunity to share the experiences of their new babies with loved ones who are elsewhere. Skype has partnered with hospitals to make Skype video calling available to patients who are separated from families.

Using video call, children can talk to family and friends and eliminate feelings of loneliness. Soldiers who are a long way from home are reading stories to their children and attending the births of their new babies.

Millions of people are connecting with loved ones from whom they are separated: parents who travel for their jobs, military families, divorced parents, and the elderly who are isolated.

VIDEO CALL CAN ALSO BE USEFUL FOR THE DEAF, HARD OF HEARING, AND SPEECH IMPAIRED, WHERE LIP READING AND SIGN LANGUAGE CAN BE SEEN.

Video calls
help you
stay in touch
with loved ones.

How to Get Started with Video Call

When I first started video calling, I used Skype, a free resource that is very popular, and this is the program/app that I continue to use today. As I began to do research for this book, I learned that there are many different ways to make face-to-face or video calls via the Internet.

This book is not just about Skype; the ideas are helpful no matter which application or program you use. Often these applications, or apps, are already installed on the computer or mobile device that you purchase; other times, you must install them. The two most popular apps are Skype and Facetime.

If you are at all uncomfortable with any of the technical terms I use in this discussion, just turn to the end of the book, where you will find a glossary containing the most commonly used computer terminology.

THE IDEAS IN THIS BOOK CAN BE USED WITH ANY APP OR PROGRAM THAT IS USED TO MAKE A VIDEO CALL.

2

Equipment You Will Need to Make a Video Call

The following equipment is required to make a video call:

1. a computer, tablet, or mobile phone that can connect to the Internet
2. a web camera (or webcam)
3. external or built-in computer speakers
4. broadband or Wi-Fi—this is how you connect to the Internet
5. a video calling application such as Skype or Facetime

THIS IS IMPORTANT—the person you want to call must have the same ***application on their computer or mobile device.*** If you are going to call family and friends, ask what app they are using and install the same application.

Skype may already be installed on your computer. If you see a Skype icon or can locate it in your programs/applications list just tap or select the app to get started. If you don't have the app you can download it to your computer

How you get ready to make your first video call will depend on how comfortable you are with using a computer. For regular computer users, it is as simple as downloading the app, following the directions on the screen to complete the set up, signing in, turning up the speakers and selecting the video call button.

For those who are not familiar with how to download programs, I have included steps to follow to be successful. If you are a visual learner for technical skills, as I am, take advantage of the videos on the Internet to view the steps in this process.

Help for Beginners

Don't be embarrassed to ask for help. If you are new to using a computer or just feel overwhelmed with the thought of technical chores, have a family member, friend, or professional technician help you get started. They can evaluate what equipment you have and what you need. Evaluating your equipment should only take a few minutes.

Family members and technicians can purchase the equipment that you need, bring the equipment to your home, and set it up. They can download Skype or another application and help you get the program working. With this assistance, all you will have to do is select a picture or symbol (called an icon) on your computer screen, and you will be ready to call—very easy.

Don't miss out on the fun because the process seems complicated; others can assist you in getting set up. Skype is available for military personnel in many locations. Hospitals, assisted living facilities, and nursing homes have computers set and ready to go—all you have to do is select the video call icon. If you are away from your own home, ask about access to video call.

Helpful and Free Videos

If you have no experience in downloading programs or applications from the Internet but would like to do this yourself, there are many great videos on YouTube that can walk you through the process of downloading and setting up your program of choice. Go to www.youtube.com and type in the search box ***How to make a video call with Skype*** or ***How to use Facetime***. You will find tutorials from the manufacturers as well as helpful videos from enthusiastic users of those and other programs.

WATCHING A VIDEO OR TWO WILL HELP YOU SEE HOW EASY SKYPE OR FACETIME CAN BE. TYPE YOUR REQUEST IN THE SEARCH BAR, FOR EXAMPLE:

- ***HOW TO VIDEO CHAT WITH SKYPE***
- ***VIDEO CALL WITH FACETIME***

Steps for Checking Your Equipment

If you are a do-it-yourself kind of person, checking your equipment is not difficult.

Step 1
Checking Your Equipment - INTERNET CONNECTION

The first thing you need is an Internet connection with broadband or cable; a dial-up connection does not work well for video calling. Verify your connection by opening your web browser and accessing the Internet.

INTERNET WEB BROWSER ICONS

CHROME EXPLORER FIREFOX NETSCAPE SAFARI

Step 2
Checking Your Equipment - WEBCAM

Determine if you have a webcam—this is the camera that is used for the video portion of your call. Most laptop computers and tablets, such as the iPad, have a built-in camera; smartphones also have cameras. A desktop computer may not have a webcam. The webcam is not part of the actual computer but rather is integrated into the monitor or computer screen. Many monitors do not have cameras, but you can purchase monitors with a camera.

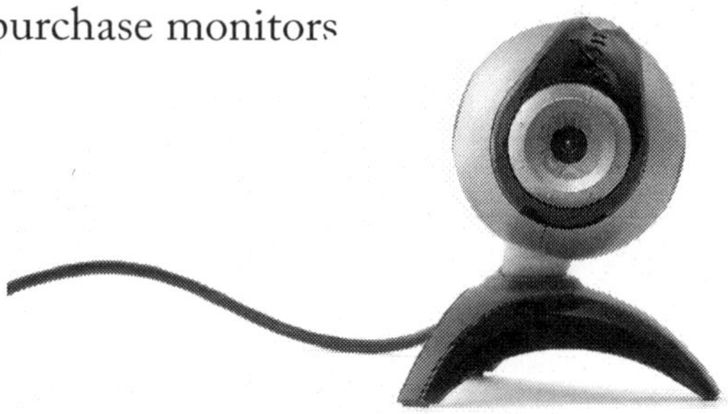

Webcam

If your computer doesn't have a webcam, you can purchase one at an electronics store or online. A webcam is placed either on top of your monitor or on your desk. Some are plugged into a USB port on your computer, others are wireless. If you have a webcam that you plug in and then disconnect between calls, you must remember to plug it back in before you call again, or you will not see a video picture when you call.

USB Port

Step 3
Checking Your Equipment - speakers

You need speakers and a microphone for sound. Most computers, tablets, and smartphones come with speakers and an internal microphone, so you may have all you require without making an additional purchase.

Sometimes speakers are present but are not connected. Check to see if you have external speakers, and be sure that they are plugged in to the computer and the volume is turned up.

For Those With Hearing Difficulty

Video call is helpful for those with hearing problems as it allows for face-to-face contact, lip reading, and sign language. The quality of sound can often be improved with a headset, especially for those with hearing problems.

3
Let's Skype

This chapter is for those who will use the Skype application. If you are using FaceTime, skip to chapter 4. Skype downloads are available at www.skype.com. To locate other video call applications, use the search engine in your web browser and type in the name of the application you want to use.

ONCE YOU DOWNLOAD THE SKYPE PROGRAM TO YOUR COMPUTER, YOU CAN ALSO CHECK YOUR EQUIPMENT TO BE SURE IT WORKS WITH SKYPE.

Step 1:
Download Skype to Your Computer

Be sure that you are connected to the official Skype website, www.skype.com to download and install the Skype app. Select the **DOWNLOAD SKYPE** button. The buttons on the Skype website change from time to time so read carefully and select the version for your computer. There are options for Windows as well as Mac and Linux users. There are also separate links for downloading to a tablet or mobile device. Get some help from family or technical support if you are unsure of which version you need for your computer.

After selecting the **DOWNLOAD SKYPE** button, follow the steps on the screen. If your computer blocks the download, you should receive a message at the top of your computer screen to see options—choose **DOWNLOAD FILE**.

The file should now download from the website to your computer. If a security warning pops up, select **RUN**. If you are asked again, it is ok to select **RUN** again for additional files. Many applications ask you to install additional features, such as search bars, to your computer. ***You do not have to agree to accept other features.***

Once you read the Skype Terms of Use and Privacy Policy, you can select **I AGREE** to continue. When downloading the Skype app you may be given the option to either accept or reject other features of Skype such as the Click to Call Feature. Microsoft may offer the option to use BING as your search engine or to change your internet home page. Read the descriptions and make a decision. To install the feature on your computer leave the box checked. To decline, tap or click on the box to remove the check

You will see a bar that shows you that the program is downloading and you should receive a message from Skype when it is complete. Usually the download process also puts a Skype icon on the desktop screen of your computer. To start using Skype, simply tap or double-click on the Skype icon. You **START** Skype as you would any other app on your computer, tablet or phone.

Mac users save the Skype file by selecting the file, then dragging the Skype icon to the Applications folder. Launch Skype from your Applications folder and sign in with your Skype name and password.

Step 2:
Create Your Skype Account, Name, and Password

After downloading the Skype app you can sign in with an existing account or create a new account. You can also sign in to Skype using login information from a Microsoft account, or by using Facebook. If you have not used Skype before you must create a new account. Select the link to Create an account. Some of the information on this page is required but most is not. Anyone on Skype can see your profile information so be safe - do not include any information that you don't want the public to have access to. You are required to provide your full name and your language preference. You are also required to create a Skype name and password.

When you create your Skype name you can use your name or some form of your name. If the name you create is already being used, Skype will let you know that it is not available, and you can choose a different name. Skype will then reply with suggestions for possible names. Create a Password and repeat the password to confirm it.

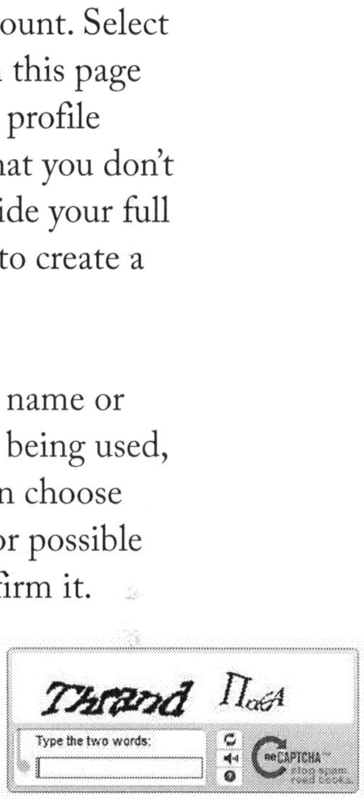

Example of a captcha security feature

You will now see a strange word or group of letters/numbers in a box at the bottom of the page. This is a security feature called a ***captcha.*** Enter the letters or words that you see in the box. Sometimes the letters are difficult to read; if you cannot read them clearly, select the **REFRESH** button to get some new words. There is also an option to listen to what is in the box.

Now read the Skype Terms of Use and select the **I AGREE** and **CONTINUE** buttons. If all is entered correctly, Skype will send you a confirmation. If not, then you will be asked to review the information that you have entered and make corrections.

Step 3: Personalize Your Skype Experience

Once your video call program is downloaded to your computer, you will have some options to consider. Select a picture that represents you that will appear on the screen of those whom you call. Select a favorite picture of yourself, of your pet, or of some familiar landmark where you live.

Be creative, have fun, and don't worry—if you don't like what you have selected, you can change it. My picture is of palm trees and the sun setting—like many grandparents, I live in Florida.

You also have an option to take a picture of yourself using Skype, and many people do that. View your public profile (what others will see) on Skype by selecting **VIEW,** then **PROFILE**. Change anything that you don't like.

For your personal icon, choose a picture of yourself or something special to you, such as your pet.

Step 4:
Add Contacts

Before you can call family and friends, you need to add their names to your **CONTACTS** list.
Select the **ADD CONTACTS** icon.

You can search for the person you want to add by his full name, Skype name, or e-mail address. Enter that information in the search box. If there is a match, the name and profile picture will appear. Select the box that says **ADD TO CONTACTS** and type a short note asking that person to Skype with you.

Skype 'Add Contact' icon

If you have any difficulty finding family or friends, ask them to send you their Skype names. They can also send you an invitation to be added to ***their*** contacts, and then you will be able to easily add them to ***your*** contacts as well.

Once the new contact is added, you will see a question mark next to the person's status until he or she accepts your invitation to Skype. When your contact accepts your invitation, you are ready to make a video call.

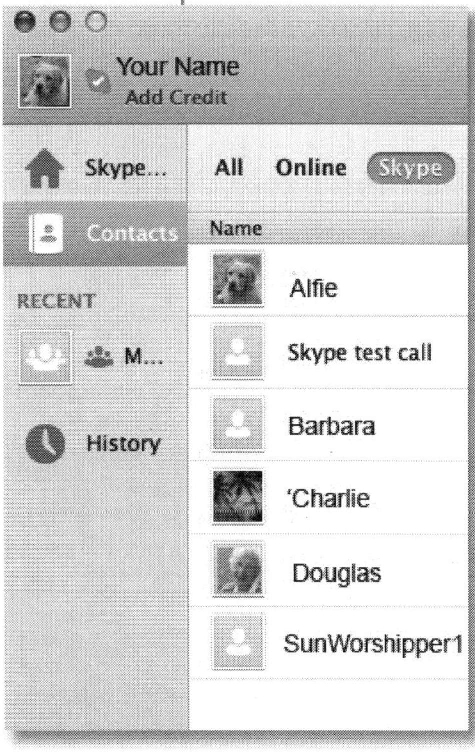

Step 5: Make a Test Call

A great feature of Skype is that you can make a test call to check your computer speakers.

In your list of **contacts** you should see **Echo/Sound Test Service** or **Skype Test Call.** If you don't see Echo/Sound Test Service or Skype Test Call, type **Echo** in the search box right above your contacts. Once you see the test, tap or double click on it and it will bring up the sound test window. Click the green 

button to start the test.

Follow the instructions and you should hear a recording asking you to record a message. Record a short message. Once you stop speaking, the test will play back your recording, and you will be able to hear the quality of the sound. If you are able to hear your own voice, you have been successful in setting up the sound equipment. You can select the red telephone handset icon to hang up the test call. There are troubleshooting videos in **SKYPE HELP**, so select **HELP** in the menu at the top of the home screen if you have problems with the test call.

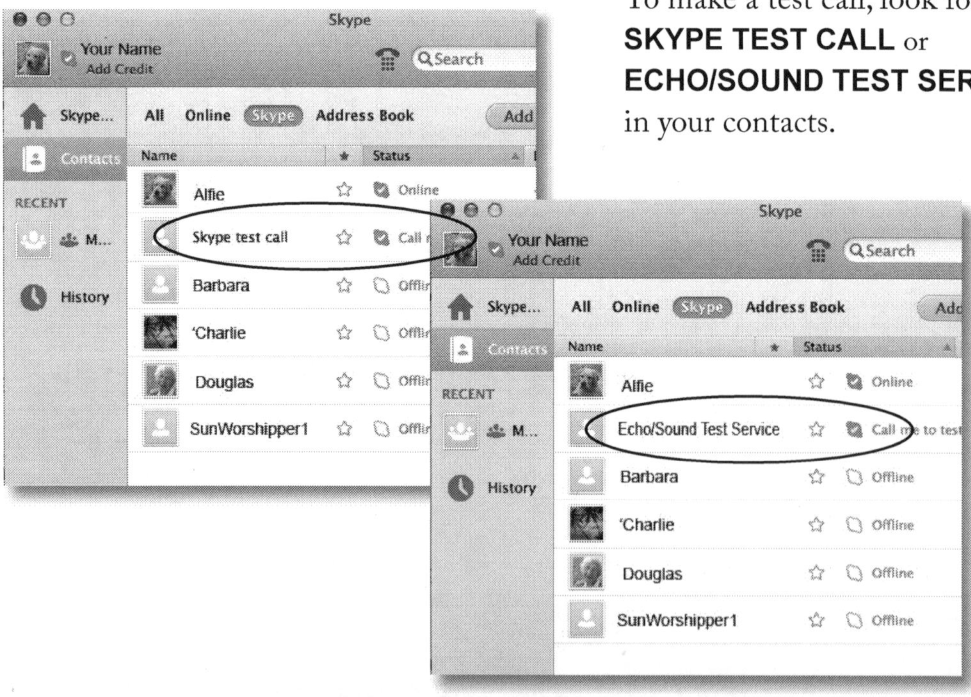

To make a test call, look for **SKYPE TEST CALL** or **ECHO/SOUND TEST SERVICE** in your contacts.

4

FaceTime for Mac

Apple's Macintosh computers offer an application called FaceTime to make video calls to other Mac computers, and to an iPad, or iPhone. FaceTime has the picture-in-picture view, as does Skype, so you can see how you look to the person you are calling. You will need an Apple ID and email address to make video calls.

All Apple products now come with the FaceTime application already installed so you will not have to download that app to your computer or phone. Downloads are available online for those with older Mac devices. If you will be calling friends and family who use Apple devices, you will most likely use FaceTime. At the time of this writing if you plan to call Windows or Android users, Skype, Yahoo Messenger or Fring are compatible applications you can use.

With FaceTime, as with Skype, you also use a **contacts list** to make the calls. Once you enter a contact into your phone or computer, it automatically shares this information across all your Apple devices. For example, to call your sister Sarah, just find her in your contacts, tap the FaceTime button and you have a video call.

With some devices you also have the option of asking Siri, Apple's intelligent personal assistant to make your call. Siri lets you use voice commands to perform many different tasks. You can make a video call by asking Siri to FaceTime your sister.

You can also switch from a regular telephone call to a video call while talking on the telephone. While talking to your sister on an iPhone you can just tap the FaceTime icon on the screen and see your sister smile.

Whether you use the FaceTime button or Siri to make your call, FaceTime will extend an invitation to the person you are calling. When that person accepts the call, the video call is initiated.

A great feature of FaceTime is that when someone calls you, that call rings through on every Apple product that you own, even if FaceTime isn't running. **There is a Do Not Disturb** function that can be activated for times when you don't want to receive calls.

FaceTime does have some limitations. Unlike Skype which can be used with the operating systems of most computers, FaceTime is only compatible with Apple devices. FaceTime is also limited to making calls over Wi-Fi or cellular, which limits where you can make calls. Group calls are also not possible at this time, although free apps like Skype or Fring can be used on Apple products to make group calls.

Lastly, permanent privacy settings are limited. You will find more detailed information in chapter six.

Most FaceTime users agree that it is very easy to use and the video quality is great, so FaceTime is a good application for video call, with some limitations.

Checking Equipment With FaceTime

Equipment

You will need an iPhone 4 or later version, iPad2 or later, or iPod touch, 4th generation or later, with the latest available version of the iOS. This is the operating system for the device.

An Apple ID and email address is required. If you plan to use FaceTime with your phone you will need an SMS messaging plan. SMS stands for Small Message System, a text messaging system. Check with your phone carrier to see if your phone has this option as well as a cellular data plan.

To use FaceTime, an active Wi-Fi connection or cellular data plan is required. Verify that your carrier supports FaceTime.

When you activate and register your Apple device it automatically registers your phone number for use with FaceTime during that process.

To use FaceTime with your Apple ID, navigate to **SETTINGS** then **FACETIME** and tap **Use your Apple ID for FaceTime.**

If don't have the latest operating system for your device you can download it at www.supportapple.com. If you have problems with connecting or using FaceTime you can find help at that same website.

5
Tips for Improving the Quality of Your Video Calls

Location

With desktop computers, you are limited to being indoors, but for many types of calls, this is not a problem. It is easier for children to focus on you and your face if there is not a lot of clutter in the area where you are sitting. With mobile telephones and many tablets, you have few limitations. You can call from the backyard, the playground, or the beach—what fun!

Lighting

Good lighting is important for the quality of your picture. A room that is too dark or too bright will reduce the quality of the picture that you want others to see. When you make a video call, you will see a picture-in-picture view. You can see a picture of yourself (what the person you are calling sees) on your screen as well as a picture of the person you are calling. Check how you look on the screen and see if the brightness is ideal. If you are indoors and your picture looks dark, add lighting, such as a lamp.

Sound

Can you hear and be heard? Be sure that your speakers are attached and the volume is turned up. If the volume is too loud, you might get a distorted sound (or feedback), so you may need to do a little adjusting of the speakers. If using Skype, test the quality of your sound before you make a call. See the section entitled **Make a Test Call** in the previous chapter.

DON'T FORGET TO TRY A HEADSET IF YOU HAVE DIFFICULTY HEARING.

Skype Help

There is excellent information on how to use Skype and how to troubleshoot problems. on the Skype website. To get to the support information select **HELP** at the top of the home screen and then select **GO TO SUPPORT**. Users can search frequently asked questions (FAQ), read the How-to-Guides and select videos to watch. There is helpful information for Windows and Mac users.

Make Your First Video Call With Skype

Sign on to your Skype account and select the **CONTACTS** tab at the top of the screen. In your list of contacts, check to be sure that the person you want to call is online. You will see the person's picture next to their name as well as their status, online, offline, or away. Select the name of the person you want to call. This will bring up the green **VIDEO CALL** button.

Now select **VIDEO CALL**. If your contact answers, you will see a live picture of her as well as a small inset picture of yourself. If the person doesn't answer, select the red telephone handset icon at the bottom of the screen; this is the end-call button.

If you are using an external webcam and do not see the **VIDEO CALL** button appear, check to be sure that:

- your webcam is plugged into the USB port on your computer
- the person that you are attempting to call is online

IF THE PERSON YOU ARE ATTEMPTING TO CALL IS NOT ONLINE, THE VIDEO CALL BUTTON WILL NOT BE VISIBLE.

Make A Group Call with Skype

While you are on a video call with any contact, you can add another contact for a group call. At one time a Premium Membership was required to make group calls. However, in 2014 group calls became free for all Skype members. At this time, group calls are not available with FaceTime.

Skype group icon

Select the group icon. A list of contacts will appear. From the contacts tab, select the contact that you want to add to your group call, and drag the contact into the empty group space. Select another contact to add, and drag the second contact to the group space. Repeat this process for all you want to include in the group. All contacts must be signed in to Skype to be added.

You can also set up a family or friends group in advance of your call by clicking on the **GROUP** icon and selecting the group members from your contacts. Skype recommends that for the best quality, add no more than five others to your call, although a total of ten users is possible.

There are alternate ways to create a group so you might take a look at different options at www.support.skype.com and select the process that works best for you.

At this time, group calls are not available with FaceTime. Owners of some Apple devices can use Skype or Fring apps to make group calls.

Video Call Using a Television

Some newer televisions make it possible to make calls right through the TV with a Wi-Fi or Ethernet connection. There is also a gadget called TelyHD that you can clamp to your TV or use on a tripod or shelf. The TelyHD has a camera, microphones, and the guts of a computer all rolled into a footlong tube. Television video call is not as commonly used as video call on a computer—it's a newer technology, but it has the advantage of a bigger screen and allows whole families or groups to fit into the picture.

TelyHD

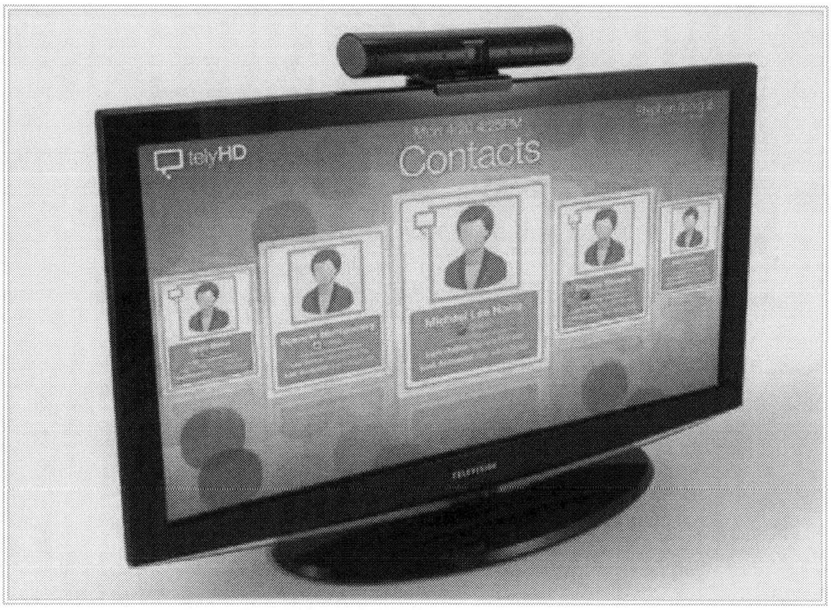

Television with TelyHD

6
Protect Your Privacy and Avoid Internet Fraud

Always think about protecting your identity and the child's identity when using the computer. Computer scams and identity theft are now widespread. Don't avoid this wonderful technology; rather, learn to use it safely.

Privacy Settings

THIS IS IMPORTANT! Before you get started using any new application, including Skype and other video calling apps, look for the **PRIVACY SETTINGS** in the application that you are using and set them to ensure that your privacy is protected. If someone else sets up your application for you, be sure you ask the person to take care of the **PRIVACY SETTINGS**, and/or **PARENTAL CONTROLS** for children.

Those who are new to computing, especially seniors and children are particularly vulnerable to scams, so don't allow calls with anyone you don't know. Don't add any information that you don't want shared with others. If you are using the app for video calling only, you don't need to fill in your telephone number, address, or birth date. The less information you share publicly, the safer you are.

Contacts and Privacy

In computer language, contacts are people whom you know and communicate with. Computers have taken the place of address books for storing this information, and that list of friends and family is called your contact list. It is a great way to organize information, but when your personal information, or that of others, is online, you need to use caution.

For those who are new to using computers, a contact list can contain names along with their associated addresses, e-mail addresses, telephone numbers, and other personal information. Skype, Facebook, and many other applications let you decide who you want to be in your contacts, just as you would choose what names you would write in a paper address book.

If you see a contact with a question mark in front of it, or a name that you don't recognize - select once on the contact to see more information about that person. In Skype, if you don't recognize the contact name after selecting on it, go to the pull-down menu at the top of your screen and select **CONTACTS** then **REMOVE**

It's important to know that you can and should pay attention to whom you allow to be a contact when you are online or using a video calling application.

If you use Apple products, choose the contact, tap **EDIT**, scroll down and tap **DELETE CONTACT.**

Protecting the Child's Safety

This wonderful video calling technology also has risks for children, so protecting your child's safety must be a priority. Using a computer in an open, common area like the family room or kitchen will make you more aware of what the child is doing.

The United States Federal Bureau of Investigation (FBI) website: **www.fbi.gov/fun-games/ kids/kids-safety** has an excellent page to help you know what limits to set and provides important information to teach children.

FBI Online Safety Tips

The computer safety tips for children on the FBI site include:

- Never give your name, home address, school name or telephone number to anyone online that you don't know
- Tell your parents right away if you read anything on the Internet that makes you uncomfortable.
- Never write to someone who has made you feel uncomfortable.
- Never meet with someone (in person or online) without your parent's permission.
- Know that people online may not be who they say they are. Someone who is an older man can pretend to be a 12-year-old girl.

Thcsc arc critical points for all arcas of computer use, not just with video calling. Parents should be sure that the settings on a computer that the child will use do not allow calls from anyone other than those on a parent-approved contact list.

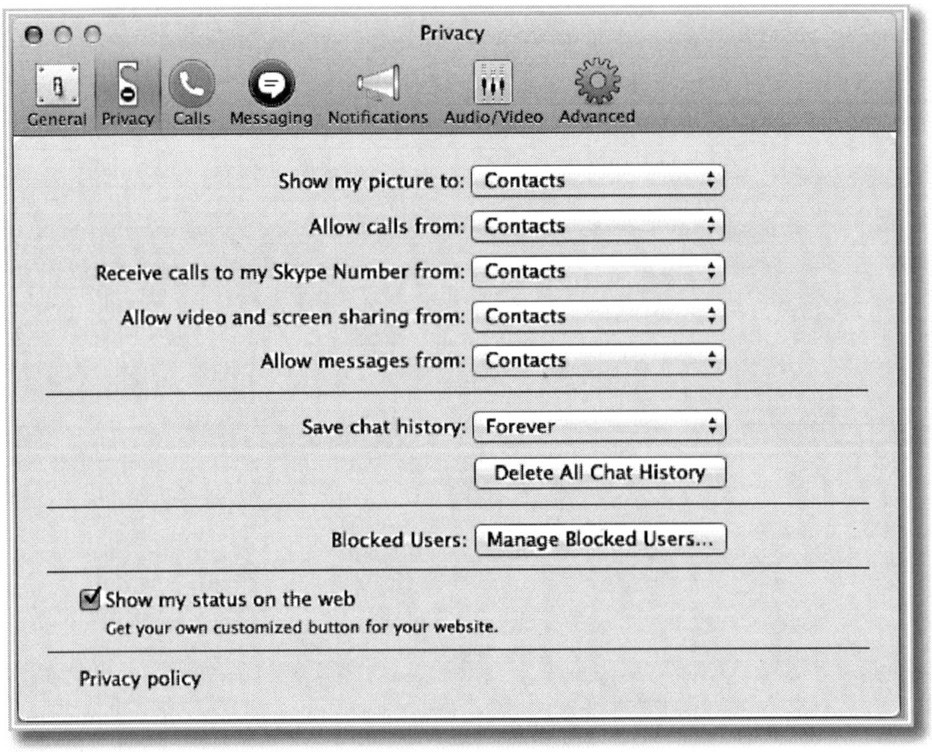

Skype privacy panel

Privacy Settings in Skype

In Skype, privacy settings are in the pull-down menus at the top of the home page. When you tap or put your mouse over the word Skype, you will see a menu of options. This is where you can fill in and edit your personal information or profile.

Don't add any information that you don't want shared with others. If you are using Skype for video call only, you don't need to fill in your telephone number, adddress or birth date. **The less information you share publicly, the safer you are.**

DISABLE THE OPTION TO ALLOW CALLS FROM OTHERS NOT ON THE PARENT-APPROVED CONTACT LIST.

When setting your privacy options, ***you*** decide from whom you will accept calls. I recommend that these settings only allow calls from people in your contact list. The same can be done for sharing videos and allowing instant messaging, or *IM*. Check that you want to allow these activities only from those people you have added to your contact list.

You can also decide if your want your online status (whether you are on the Internet) to be visible on the web each time you sign on. Uncheck this option if you don't want to appear online once you finish making or receiving a call.

Skypito

Designed by EasyBits in conjunction with Skype, Skypito is an application you can download free of charge. It is designed for children two to fourteen years of age to help them feel independent while using the computer, at the same time giving parents control of the contact list.

Parents preapprove, with a password, those individuals with whom a child is allowed to chat. The child friendly graphics and games make this program kid friendly. It requires the regular Skype program to be downloaded as well.

Explore Skypito at www.skypito.com. Don't let all this important privacy information prevent you from using video calling. You are in control, and video calling is too good to miss. Just as there are telephone scams, there are also online scams—be aware and take precautions.

Webcams

It is possible, although not a common problem, for hackers to take control of your webcam. Some security experts suggest that you disable the webcam using the system settings while others suggest that you just cover the camera lens with a small piece of opaque tape when you are not using it. If you want to use a video call app often, covering the camera when not in use is the simplest solution.

Parental Controls in FaceTime

At the time of this writing, Parental Controls on Apple devices allow you to restrict the use of FaceTime but not the ability to specifically restrict calls from those who are not in your contact list. Anyone who obtains the child's email address could make a video call. To control use, parents must be present during calls or restrict the FaceTime program. If you have reason for the child to use a video call program without supervision, for example with latchkey children, you may want to install Skype to set specific privacy settings.

The process for disabling FaceTime may vary slightly from one device to another, but usually it works as follows:

- Tap settings
- Tap General
- Tap Restrictions
- Enter a pass code
- A list of applications that can be enabled or restricted appears
- You can restrict access to FaceTime

Young children should not use FaceTime without parental supervision. To ensure safe use, FaceTime should be disabled when the parent is not available and then enabled for supervised use.

Depending on the age of your child you may want to be certain that he cannot guess the password for switching applications on and off. You may also want to change the password from time to time.

Part II
Let's Video Call with Kids

7

Child Development and Video Call

Milestones

Knowing some basic information about what children can do at each stage of development helps parents and grandparents know what the child will be able to respond to when video calling.

Penelope Leach, in *The Essential First Year*, describes milestones like rolling over, sitting, crawling, and standing as useful reminders of what we can expect from babies, but she reminds us that they tell us little about exactly when we can expect it.

> THERE ARE MANY VARIATIONS IN NORMAL DEVELOPMENT.

No two children develop in exactly the same way, so keep in mind that the descriptions presented here are very general. Some children will master skills earlier and others later. Two children the same age playing together often have different skills, so don't compare your child or grandchild with other children the same age. Parents should seek advice from the child's health care provider if they have concerns.

Early development is important for a child's health. Information about milestones is easy to obtain in books and on the Internet. The Centers for Disease Control and Prevention (CDC) is the national public health institute for the United States and has an excellent website that provides information about child growth and development.

Websites to Learn About Child Development

- www.cdc.gov/ncbddd/child
- www.babycenter.com
- www.scholastic.com
- www.pbs.org/parents/child development

It is helpful to know in advance from the parent what skill has been mastered so you can ask a related question and the child can have fun demonstrating for you.

For example, at age two, my grandson was playing with matchbox cars, and when I asked, "What color is this car?" he responded with names of colors, but they were not correct. I decided to check to see when learning colors commonly occurs with toddlers. I found that most toddlers can name one color at thirty-six months (remember that some will do this before thirty-six months, and some after).

After understanding that he had not reached that age or developmental stage, I named the colors for him for several weeks rather than making a game of him naming colors, and we enjoyed our video time together. It wasn't long before he was naming the colors himself.

Be aware that preemies (children born before full term) may be slower in demonstrating the skills that are mentioned in this book. Some mothers of preemies deduct the months that the baby is early when looking at what the baby can accomplish at each developmental stage. This correction is generally made until the age of two years; eventually, preemies catch up.

General Guidelines

It will take a few calls to get a sense of what works best for you and for the child, but keep the following in mind:

- Attention spans are short, so keep the call short or make attempts to engage young children for short lengths of time.
- Be flexible; most young children will lose interest in speaking after a short time. Sitting back and watching a child play is just as enjoyable as trying to engage in conversation once the child begins to lose interest.
- When children are tired, hungry, or frustrated about something, don't expect them to enjoy chatting, especially between the ages of two and three years.
- Not all babies and children enjoy the same activities, so some may respond happily to my suggestions while others may have little interest. Use the ideas that work for you and the child you are calling.
- Once a child moves from one age to another, don't hesitate to go back to some suggestions for activities listed for younger children. Children love repetition and returning to favorite games and songs.

In the next section, I share some ideas about toys and activities with which the parent with the baby or the video caller might enjoy experimenting. These are just that—suggestions. Don't be afraid to try playing peek-a-boo or singing songs. You will be pleasantly surprised to find that the child is more engaged and you are having fun, at the same time as you are building a wonderful, lifetime relationship.

Dorothy Einon has written several books on play with suggestions for toys by age. Einon's *Play with a Purpose* and *How Children Learn through Play* may be of interest to you.

Children of all ages enjoy books. Resources that provide ideas about which books are best for different ages are available.

Consult the experts at your local library and bookstores, or use online resources such as www.Scholastic.com and www.Amazon.com. Pam Allyn's book *What to Read When* is a great resource for parents.

Birth through One Month: Connecting

Video calling with a newborn to one-month-old baby is a wondrous experience. You begin this new relationship by connecting with sound. Baby's distance vision is limited and will take several months to develop. However, the infant's sense of hearing is well developed as the baby has been listening to sounds inside the womb for many months. Parents' voices are recognized, as are the voices of others who are frequently around the baby or who were often with the pregnant mother.

In the Yale Child Study Center's *Guide to Understanding Your Child,* the authors tell us how important sound is to babies. After birth, a baby is more likely to recognize his father's voice if he talked to the baby while it was still in the womb.

Parents, grandparents, aunts, uncles, and siblings can begin a relationship with video calling so that the baby will also recognize their voices, even if they are not with the baby frequently. High tones and gentle sounds are preferred–avoid deep and loud sounds, which may frighten the baby. Adults often use high tones when talking to babies, so a little gentle talk from you will sound just right at this early stage.

Because the baby's skin is sensitive, the person who is with the baby should hold and stroke the baby during the video calling session so that the baby connects a loving touch with the sound of the voice.

What to Have Ready for the Video Call - BIRTH THROUGH ONE MONTH

These first months are so easy. No special equipment or toys are needed. Mom or Dad and baby should find a comfortable position in which to rest and relax during the video visit. Sometimes parents can plan the video call during times when baby is active and alert, but with babies and young children schedules can change.

Video calling at this point is enjoyed by the proud parents and brings joy to grandparents and other family members who may not be geographically close enough to see baby on a regular basis. Just remember that parents are often tired, so let them determine the best time and length for a video visit— short visits are often best received.

Keep the call short and be flexible when change is needed.

What to Watch For - BIRTH THROUGH ONE MONTH

Watch as baby begins to stretch her arms and legs. The motion you will see is mainly the result of reflexes or automatic responses such as sucking or closing the eyes to bright light. A newborn baby doesn't have control over her body and cannot manipulate playthings. Movement will be jerky. Hands are usually clenched; baby can't control her hands but may begin to watch them. Baby's sight is not clear; she sees only a short distance and will not see you for a while.

Hearing your voice is the best introduction at this point. Newborns have good hearing, so talking, singing, or playing music is stimulating. The lighting should be indirect if you want to see baby's eyes open. If you want to see her move her eyes and head, toward the end of the first month, have Mom hold an object eight to twelve inches away from baby and move it slowly from one side to the other. Baby's eyes and head may follow the object.

The Second Month

What to Have Ready for the Video Call

✤ Rattles or small toys that the baby can follow with his eyes
✤ The words to some of your favorite children's songs
✤ A music box or instrument, a recorder, or an iPod with music

What to Watch For - THE SECOND MONTH

Baby's eyes can move and focus, but focusing on the computer screen for more than a short time is unlikely. Baby can see you from a semi-sitting position and can listen and make sounds. He will enjoy listening to you sing, so this is a great time for you to sing some of your favorite children's songs: *Twinkle, Twinkle Little Star, The Farmer in the Dell, Row, Row, Row Your Boat.* See the list of Favorite Children's Songs for ideas. If you can't remember the words to the song, you can easily find them on the Internet—just type the title of the song into Google or another search engine. If you are not comfortable singing, use a music box, musical instrument, CD player, or iPod with children's songs.

Baby is also enjoying the sound of his own voice, so try playing the echo game by repeating baby's sounds — and watch for early smiles.

The Third and Fourth Months

What to Have Ready for the Video Call

- ✤ Objects that have bright contrasting colors such as yellow and dark blue or black and white
- ✤ The words to some of your favorite children's songs
- ✤ Small finger or hand puppets
- ✤ A music box or instrument, a recorder with music
- ✤ Books with big, bright pictures

What to Watch For - THE THIRD AND FOURTH MONTH

Hello out there! Baby can now focus on other people besides her parents, and her distance vision is improving. Baby might also notice her own picture on the computer screen and smile at the baby! She is beginning to develop some control over her hands, arms, and legs. By this point, baby will begin to smile to get your attention, and the smile may be followed by some babbling to really engage you. By responding to baby, you will let her know that she is important to you, and your responses can help build her self-esteem.

Baby still can't hold up her head for long periods at this age, so a reclining car seat, carriage seat, or Mom's lap is a good spot from which baby can see the computer screen.

She prefers bright or intense colors, bold patterns, and black and white. She can look sideways while on her back or pick her head up while on her tummy. Parents can change baby's position several times while video calling to keep baby's attention and to provide an opportunity for you to see all that she can do.

Playing games with touch and sound, such as ***Itsy Bitsy Spider,*** can be fun for you as well as delight baby. A parent can play the game by moving his or her fingers up baby's arm while you sing the song. With another favorite, pat-a-cake, you can sing the song while the parent pats baby's hands together. Changing position may engage baby for a longer period. Have the parent place baby on his stomach, and use a small puppet for a short puppet show.

Listening to stories or sharing books helps to foster language development. Babies who hear stories learn to speak more easily. Don't forget that short attention span and select a board book with bright pictures and few words. You will need to hold the page close to the webcam so it can be seen.

If you feel awkward or silly talking to babies, start with simply telling baby what you did during the day—went out to work, went to the grocery store, played golf, swept the floor. Babies and young children love sounds, so make the sound of the broom, growl like an animal, make car sounds. We are not talking about an intellectual conversation here, and it may feel silly, but children love sounds. Just mentioning a bear to my grandson produces a loud growl and a big smile.

A baby swing or bouncer is another seat that can give baby a good position for viewing the computer screen; he may not stay still for long, but you will enjoy seeing him move.

The Fifth, Sixth, and Seventh Months

What to Have Ready for the Video Call

- A scarf or towel for you
- Unbreakable blocks or cups that you can stack or baby can stack with a parent
- Children's instruments such as maracas, a xylophone, or a small drum
- A ball or a shape sorter
- Picture books or magazines with bright pictures

What to Watch For - THE FIFTH, SIXTH & SEVENTH MONTH

Baby wants to participate. Try peek-a-boo—you can use a scarf or towel or just your hands. If you say ***peek*** or ***boo*** with a smile in your voice, you will likely get a delightful reaction. Vision is improving so facial gestures and enthusiasm go a long way to delight babies.

Use simple words–*hi, mama, dad, dog, cat, bird*—and listen for babbling sounds in return.

Ask questions: "Where is Charlie (the dog)?"

Also be sure to use baby's name: "Would Joseph like me to read a book?"

Have fun stacking blocks, and then, when you knock them down, laugh–baby might surprise you with a laugh in return. Babies also delight in banging toys and making a banging sound. Blocks, stacking cups, and musical instruments are all enjoyed, and you can sit back and watch the fun. Ask, "What happened to those blocks, Brandt?"

Fine motor skills are developing, and the banging noise is an incentive for baby to use his hands. A rolling ball is also very tempting for baby to watch, push, or chase, especially if baby has begun to crawl.

Physical coordination is improving and you may be able to see baby discover and grab feet and toes and put them in his mouth.

In the second half of the first year, several new motor skills are attempted and mastered, such as sitting, crawling, and standing. There is wide variation in the age when crawling first appears.

By the sixth month, vision is well established, and many babies are able to sit with aid. This creates a shift in how baby can view the computer screen.

Continue to sing children's songs and try the scales—Do, Re, Mi, Fa, Sol, La, Ti, Do. Watch and encourage baby to put objects in a shape sorter and poke at shapes with little fingers.

Eight to Twelve Months

What to Have Ready for the Video Call

- Picture books
- Hand puppets
- Peek-a-boo scarf or towel
- Boxes, bowls, metal pans (can be a little noisy but fun), large spoons, paper tubes, egg cartons, balls
- Cars, trucks, or airplanes
- Washable markers or crayons

What to Watch For - EIGHT TO TWELVE MONTHS

Now the fun really begins - baby is much more mobil. When you video call, you will see obvious and delightful developmental changes from week to week. You can see baby's first attempts to get to a sitting position and early attempts at crawling or creeping on hands and knees. With parents' permission, call at mealtime and watch baby eat, or if baby loves her bath, call at bath time.

Baby can probably sit without support but many topple over while you watch. Be careful not to appear upset as he will soon figure out how to get back to a sitting position. You may see him reach out for toys he enjoys.

Soon you will see your little one pulling himself up to stand and walk while holding on to furniture. How exciting to see baby crawling toward the computer screen and grabbing on to stand up and see you better. Waving hello or good-bye can be demonstrated and might be accomplished between months nine and twelve.

- ✤ A portable computer on a table aimed at the baby sitting in the high chair gives you and baby a great view of each other.
- ✤ Pointing the webcam down helps you to see crawling and playing on the floor.
- ✤ Repeat simple words that baby knows, such as dad, mom, uh-oh, or bye-bye. Even if speech sounds more like gibberish at this stage, babies love to hear you repeat their sounds.
- ✤ Play pat-a-cake, peek-a-boo, and so-big.
- ✤ Play where-is and ask "where is your doll?" or "where is your truck?" and watch the child point, crawl, or walk toward the object.
- ✤ Baby can draw a simple picture with markers or crayons and show off his picture– often this is just lines.

Twelve Months to Two Years

What to Have Ready for the Video Call

- Pull toys
- Wooden puzzle with large pieces
- Stacking cups and blocks
- Toys with dials, buttons, and keys
- Children's instruments, such as maracas, a xylophone, or a small drum
- A ball or a shape sorter
- Picture books or magazines with bright pictures
- A favorite toy that was seen at Grandma's house
- Toy telephone
- Short stories to read
- Drawings or other artwork the child has done
- Child's comb, hairbrush or toothbrush

What to Watch For - TWELVE MONTHS TO TWO YEARS

As the child becomes more mobile, more objects are reachable and of interest. Most toddlers can get into a sitting position, clap hands, say one word, and pull up to standing. Many can stand without help by thirteen months, take a few steps, and point to something they want.

During this second year, toddling becomes a more mature form of walking. From thirteen to fourteen months, some will be able to walk well, pull a toy while walking, and pretend to be talking on a telephone. A few may be able to climb steps and match shapes.

By sixteen months, many will be able to stack three blocks, throw a ball, and run. A few will be able to kick a ball, brush their teeth, and say fifteen words or more. Watch their delight when they push buttons on toys that have parts that pop up. Many love to demonstrate their skills in combing their hair or brushing their teeth. Watch as the child holds a toy phone or pretends to cook, eat or drink.

Language and comprehension skills are changing rapidly. Toddlers become more responsive. You no longer need baby talk or high-pitched sounds to get his attention. By eighteen months some children can say their own name. You can teach correct names and have him point to ears, hand, toes, etc. Imitation is part of the learning process.

During this period from twelve months to two years, children may begin to repeat words that they hear, so take care what you say when calling.

When our grandson visited at twelve months, one of his favorite toys at grandma's house was a jack-in-the-box. Sharing the toy again during a video call visit brought back delighted laughs and memories.

Two to Three Years

What to Have Ready for the Video Call

- Pull toys, cars and trucks, dolls
- Balls, blocks, large LEGOs
- Action figures, people, animals including dinosaurs. For safety, use age-appropriate sizes
- Old clothes, hats, shoes, and costumes
- Colorful alphabet picture books
- Cash registers, stethoscopes, dining sets, imitation food
- For outdoors: large three to four-inch wide paint brush and pail of water, watering can, small bicycle
- New songs, for example, ***I'm a Little Tea Pot, Jack Be Nimble, Humpty Dumpty and Happy Birthday***

What to Watch For – TWO TO THREE YEARS

Preschool children are on the go, running, climbing and jumping. Their attention span is often short. At this stage rather than expecting the child to stay in front of the web cam, try having the parent direct the camera to the area where the child is playing. This position will allow you to watch the child move around the room, select toys, and return to the computer for a close up when he is ready.

> AS THE CHILD LEARNS TO SAY *HAPPY BIRTHDAY*, WITH A LITTLE PREPARATION FROM A PARENT, THERE WILL BE EXCITEMENT ON BOTH ENDS WHEN THESE SPECIAL WISHES ARE SHARED DURING A VIDEO CALL.

The years from two to three are a time of rapid language growth. Two year olds understand most of what you say to them. Many twos have a good vocabulary and start forming sentences. By age three, language development makes communicating fun. Descriptions are now in sentences. Sometimes emotional reactions are not what you might predict, so do not take it personally if occasionally your call doesn't go as well as you expected.

Most children will begin to recognizing some letters between the ages of two and three. Have fun sharing alphabet picture books – but don't expect full mastery for some time.

As the child learns to say "happy birthday," with a little preparation from a parent, there will be excitement on both ends when these special wishes are shared during a video call.

A favorite Skype activity at this age is to show the caller favorite toys and puzzles.

Children between two and three are mainly concerned about their own needs and desires. This is the stage of the ***terrible twos*** which often extends through the threes. Temper tantrums can happen during video calls, especially if the child does not want to share his parent with you, at the time of your call. Children in this age group display a broad range of emotion which you can see during a video call, as you would with any other face-to-face visit.

If you see the child begin to get frustrated or restless, do something silly or distract him with a song you can sing together. By keeping the video calls short and interesting you will be less likely to see frustration build and tantrums occur.

On some days a child in this age group may decide that they don't want to participate in a video call at all. At these times it may be helpful to let him know, prior to a call, that Daddy or Grandma has a special book to read or instrument to play for him.

Don't forget that "no" is often a favorite word at this stage. There are times when our grandchildren have not wanted to make a video call. On those days, we just enjoyed chatting with our adult children and watching the little ones play with their toys.

In the ages from two to three you may enjoy all of the following:

✤ Watching the child draw a simple picture with markers or crayons and sharing her picture.

✤ Playing a game of peek-a-boo.

✤ Watching the child paint pictures with a paintbrush or with water outdoors on a walk or driveway.

✤ Seeing her run to the computer to say hello, wave good-bye, or blow kisses.

✤ Helping a two-year-olds identify their feelings, and three-year-olds to recognize what causes their feelings.

✤ Watching imaginative play with old clothes, pots and pans and other props.

✤ Using a tablet, or phone with Skype or FaceTime to watch the child water plants, play in a sandbox, splash in a pool, and ride a small toy with wheels.

Four to Five Years

What to Have Ready for the Video Call

- ✤ Books continue to be enjoyed
- ✤ Building blocks such as LEGOs or other construction toys
- ✤ Stories to share about what happened during your day
- ✤ Objects that relate to the child's interests, such as dinosaurs or dolls
- ✤ Paper to demonstrate how to make a hat or airplane
- ✤ Materials to make a simple rain gage
- ✤ Simple jigsaw puzzles
- ✤ Many of the toys and household objects from earlier stages are still enjoyed

What to Watch For - FOUR TO FIVE YEARS

This stage of play focuses on social and emotional skills and building something while playing. Four-year-olds are more likely to initiate conversations and share personal experiences. Four- and five-year-olds continue to realize what causes feelings and understand that others don't always react as they do.

Ask open-ended questions that will most likely be answered with more than just a yes or no response: "What was the best thing that happened at school today?" or "Tell me about your best friend."

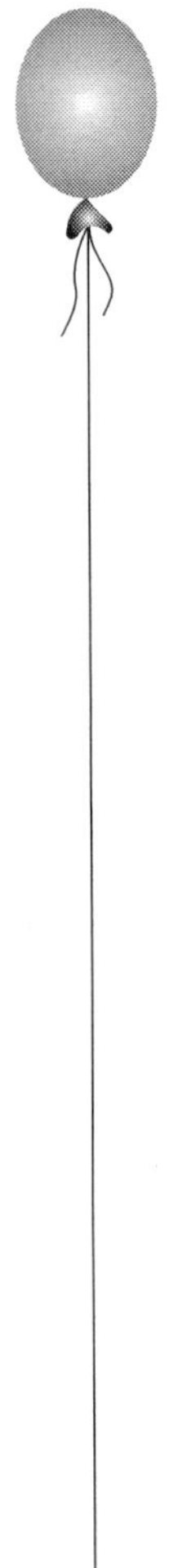

Many children enjoy making a definite structure when they build with blocks. Quality blocks are a good gift to give and can stimulate imaginative play.

If you have established a routine of reading to the child, this can be a good time to share new books. Both fiction and nonfiction books are well received. Send magazine or newspaper articles to the child and discuss during a video call.

Pretend play is still important and will become creative. Children at this age aren't always able to tell the difference between what is real and what is make-believe. At this stage, children like to show off their new skills. Watch them stand on one foot, hop, skip, and dance.

- Four and five-year-olds like to sing, dance, and act.
- The child may be able to draw a person with a body.
- She may sing songs such as *Itsy Bitsy Spider*.
- May recall part of a story or tell a story, or use her imagination to pretend.
- Can name some numbers and colors and has a sense of time.

Ages Six to Ten: The School Years

What to Have Ready for the Video Call

- Building blocks
- Send scrapbooks and journals to share
- Maps
- Collections such as stamps, sports cards, coins
- Model cars, dolls
- A single coin to play heads or tails
- Printable activities from websites
- Objects from nature, such as leaves, rocks and flowers

What to Watch For - AGES SIX TO TEN

During the early school years, children develop their motor skills, and problem-solving abilities. They develop greater independence from parents, and they form friendships. Children learn how to handle their feelings and resolve conflicts during this period. The pressures of peer relationships, school responsibilities and performance grading all come into play. School-age children may want to share their concerns during a video call.

Children enjoy collections, and you can share any that you have, encourage the child, and contribute to any of his collections. With video call, you can play simple guessing games. Games enjoyed include using a coin to play heads or tails, or your hands to play odds and evens or paper, stone and scissors. Share maps with the child to see the distance between where each of you lives, to plan trips and to identify where world events are occurring.

School-age children also love to be read to as well as to share their skills when they learn to read. Because their attention span is increasing, you can select a book to finish in one call or continue a longer book with several video visits.

What to Watch For - AGES SIX TO TEN (CONTINUED)

Being able to talk comfortably face-to-face with a child you love but who is not geographically close is a gift that modern technology has given to us.

Through the Internet, we have access to thousands of ideas about how to communicate more effectively with children and how to address tough topics such as safety, death, bullying, friendships, sexuality, and divorce.

Using the section **How to Ask Questions That Get More Than a Yes or No Answer** should help you get started in communicating with the school-age child. Activities in the **Fun Activities for Video Calls** section are relevant for a wide number of age groups, including school-age children. Parents are a good resource about what the child's current interests and activities are, so you can plan for discussion in those areas.

Adults who did not have positive role models or experiences in their own families and want to do better with their own children can learn by reading. They can discover how to interact with their children in a loving and caring way and can certainly do that using video call.

Many excellent books and online resources for parents describe the growth and development that takes place during the school years—check out some of the resources at the end of this book. If you struggle at all with how to give praise effectively, the Australian Parenting Network has an excellent video that can help. www.RaisingChildren. net.au/articles/praise_video. The Public Broadcasting Service website www.pbs.org/parents also has suggestions for parents.

Caring adults have always played an important role in a child's growth and development. The security that a child feels with the adults in his life contributes to his ability to develop positive social relationships with others. Parents and grandparents can use video call visits to provide unconditional love and a safe harbor for discussing fears and sharing accomplishments.

> "As a general principle, always keep the scales of praise and criticism firmly weighed in the direction of praise."
> *Dorothy Einon*

Parents who are away from home for work, in the military, because of divorce, or for many other reasons still have an opportunity to help their children explore problems, find solutions and manage frustrations. Discussions when calling school-age children can encourage positive actions and develop a child's sense of self-esteem. Look for opportunities to use praise and applaud efforts.

Video calls help families stay in touch

Ask Questions That Get More

Learn to ask questions that will encourage more than just a yes or no answer. In our family, the favorite dinnertime question was, "What was the best thing that happened today?"

Following are some basic questions that might help you get started during a video call:

? What was the BEST THING that happened today?

? What was the WORST THING that happened?

? What do you like most about school right now?

? I heard that you had A GREAT DAY in school today — can you tell me about it?

? I heard that you had a rough day—can you tell me about it?

? What do you like least about school?

2 + 3 = ???

? How is MATH going now?

? Can you tell me what you know about that?

? What is your FAVORITE FOOD?

Than A Yes or No Answer

? What do you like to **DO AFTER SCHOOL?**

? What would you like to talk about?

? What do you think you should have done?

? What will you do next time?

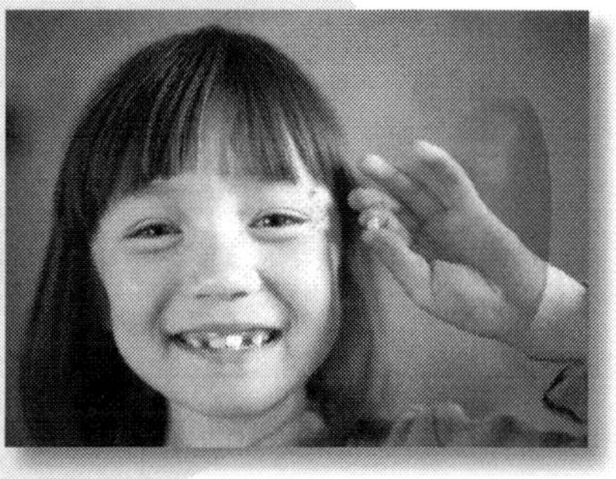

? I see you have a new haircut, **LOST A TOOTH** . . .

? What happened next?

? Can you tell me about **YOUR DOG?**

? And then what happened?

? How did that make you feel?

? I feel sad about that; how do you feel?

8

Special Considerations

Latchkey Children

Being able to see a parent or grandparent when a child is home alone can make everyone feel more comfortable. This time can also be a special time for grandparents to enjoy the excitement as the child shares news about her day or concerns she may have.

Not all grandparents can assist with homework, but some can, and with video call, many more can now provide some emotional support to the child who is home alone.

If children will be using the Internet without adult supervision, be sure that the privacy settings set the limits mentioned earlier. Consider programs like Skypito that make the Skype program child friendly.

Divorced Families and Video Call

Divorce is a process where marriages end, and custody for children can become a contentious issue. In the best of situations parents will focus on what is best for the child. Many divorced individuals as well as their children feel isolated following divorce. Healing takes many different forms in different families.

When an adversarial approach to custody is taken, children suffer. Research shows that after divorce, children do best when they have contact with both parents.

Peter Favaro, in his book *Smart Parenting During and After Divorce*, discusses the importance of making sure children stay in touch with both parents. His tips on how to avoid conflict through good communication are very helpful as are his suggestions for co-parenting at different developmental stages.

Divorce distress can vary with the age of the child and it is impossible to predict specifically how divorce will affect a child. Babies can feel the tension and stress in the household but cannot understand it. They may become irritable and cling to one parent.

Research shows that after divorce, children do best when they have contact with both parents.

Toddlers and preschoolers may become dependent and revert to more infantile behavior. They may feel anxious about the future and have nightmares. School age children, age's six to eleven, may not fully understand divorce and worry about being abandoned.

From ages eight to 11 children may blame one parent for the separation. Upset stomachs, headaches or other stress related symptoms may be present.

A primary relationship with both parents can help to alleviate the child's feelings of self-blame, loss, confusion, and sadness. Children need to hear and see both their parents, even when one parent has less ease and fewer parenting skills than the other.

Video calls are a great way to connect when parents are separated from their children. However, like any other visitation, planning in advance is important. Parents should discuss when calls are best received, and when they are not welcomed.

Birthdays, holidays and special occasions can become a contentious point when visitation is not planned in advance.

If you want to make a video call, advanced planning with your ex-partner is important.

Electronic Communication and Virtual Visitation Legislation

Some parents need assistance to stay in touch with their children. In 2004 Utah became the first state to pass legislation for virtual visitation for parents. Michael Gough, a computer security expert and advocate for automatic options for virtual visitation, was instrumental in getting legislation passed. As a divorced father he wanted to stay regularly involved and available to his daughter. This became a challenge when his wife relocated over 1,400 miles away.

There are currently electronic communication laws in several states and pre-draft bills in many other states. Gough continues to advocate for legislation for automatic options for virtual visitation and has a helpful website for parents going through a divorce. You can get up to date information about legislation and tips on the legal process for electronic communication at Gough's website internetvisitation.com. Consult with your attorney about proceeding with a workable electronic communication schedule. Use video call apps like Skype, and FaceTime to help your child stay connected to both parents.

Respectful communication with any of our new technology, whether it is mobile phones, texting, or video call is crucial for the child's welfare.

I found many tips for divorced parents on the Australian Parenting Network site **www.raisingchildren.net.au**, and at the Parenting network site **www.parents.com**.

Tips for Communicating with Video Call in Divorced Families

- Discuss in advance how the family will use the technology of video call
- Avoid using barbs or criticism; speak positively of the other parent when calling
- Don't ask the child to choose sides or which parent he likes best
- Continue to encourage the behaviors and limits that both parents have agreed upon
- Give the child praise for positive ways that they are coping with the situation
- Don't worry the child with problems parents must solve – keep grown up problems for grownups.
- Avoid asking questions about what the ex-partner is doing
- Avoid questions about the ex-partner's spouse or partner
- Make attempts to keep in touch with extended family
- Create an atmosphere where the child feels safe to discuss their thoughts and feelings
- Don't make promises that you can't keep
- Don't ask your child to keep secrets

With the wonderful technology of video calling, when you ask, “What is your favorite thing to do?” children can show you as well as tell you.

9

Fun Activities for Video Calls

Do you always have to plan activities when video calling? Absolutely not. Some of the best times happen spontaneously with no plan at all. Although this book presents many suggestions, there are times when doing what you might do if you were visiting the child at home works well. Say hello and then just sit back, relax, and chat with the parents and the child, or watch the child play with her favorite toys.

Left on their own to play, most children will bring their favorite toys to the computer screen and enjoy your comments. My grandson likes to bring toy trucks and name them for us. He loves to see our faces light up when he identifies the object, color, or shape that he is learning and enjoys sharing his new skills with us.

You can stimulate both conversation and learning by asking questions such as, What color is that? How many can you count? What shape is that? What letter does that begin with?

This chapter provides many additional suggestions.

Show and Tell

Children of all ages enjoy showing us what they found on a walk, artwork from day care or school, favorite toys and books, and new clothes. Telling us about their day is fun.

They love to demonstrate new skills and abilities. When they have learned a new skill or have progressed to a new level of accomplishment, they might want to show you how they:

- **say a new word**
- **throw a ball**
- **read**
- **skip**
- **jump rope**
- **play with a yo yo**
- **play an instrument**
- **ride a bike**
- **swim**
- **make a snowball**
- **draw or paint**
- **play with a slinky**

Repetition

Jill Stamm, co-founder of New Directions Institute for Brain Development has explored how children's brains develop. In her book *Bright from the Start*, she talks about the importance of repetition for learning and the role repetition plays in making strong connections in the brain. Repetitive movements, such as pushing, pulling, touching, and turning, help a child figure out how things work.

When a child feels safe, this learning is fun, and when activities are fun, practice improves performance. Repeat activities such as songs, reading favorite books, and puppet play. Make your time with video calling fun, and don't hesitate to repeat activities that were enjoyed at earlier stages as children grow.

Reading as Play

When you watch readers who know how to relate to children, you can see how much fun it is to read and be read to. Be enthusiastic about the story and use different voices for different characters or props to sustain interest. Hold the book page close to the video cam after reading so the child can clearly see the pictures.

Just demonstrating your enthusiasm for reading can encourage the child to read and love books. Read poems, stories, and newspaper or magazine articles. Take a look at Pam Allyn's *What to Read When* for an extensive annotated list of more than three hundred book titles for children from birth to ten years.

Sharing Jokes and Riddles

Simple jokes are a wonderful way to get your grandchild giggling.

✤ Why does a flamingo lift up one leg?
Because if he lifted up both legs, he would fall down!

✤ Why do birds fly south in the winter?
Because it's too far to walk!

✤ What is a volcano?
A mountain with hiccups!

✤ Knock, knock.
Who's there?
Olive!
Olive who?
Olive you!

Both the library and the Internet are great resources for riddles and jokes.

Playing Dress-Up

Young children love to pretend and dress up in clothes that are for adults. Send some of your old dresses, shirts, hats, and shoes and watch the fun. Make or purchase capes and enjoy watching your superhero fly around the room. Recycle old Halloween costumes or visit party stores and discount stores to find inexpensive costumes that can be used for play. In addition to local stores, several online costume resources are available, such as www.costumes4less.com.

Cooking

With portable computers that can be placed in the kitchen, you can create opportunities to share recipes and for your child to watch you cook. This can be even more fun when you make something you can send to the child such a herbed popcorn, granola bars, fruit breads, brownies, or cookies.

Opening Gifts

It's so much fun to see a child open a gift that you have sent for a special occasion. It gives younger children a real identification with the person who purchased the gift. Seeing the true reaction is a treat–that is, most of the time. I once sent pajamas that I thought my two-year-old grandson would love. What I learned was that the stickers on the package were much more fascinating than the PJs.

Enjoying Music

A good-quality xylophone is a terrific instrument for children aged three to four years. Simple, homemade musical jars (several similar jars filled with different levels of water) and played with different 'drumsticks' (wooden or metal spoons) are fun to share while video calling, and the child can make his own at home. Children who are taking music lessons can share a new song they have learned or play a favorite song of theirs or yours. The adult can also share her favorite music. If both you and the child play an instrument, try playing a song together.

Favorite Children's Songs

If you can't remember the words to any of these songs, type the title into your Internet browser. The words to all these songs are available on the Internet.

A Frog Went a Courtin'
A Hunting We Will Go
A Ring a Roses
A Tisket A Tasket
Alphabet Song
Baa Baa Black Sheep
Baby Bumble Bee
Bear Went Over the Mountain
Bicycle Built for Two
Bingo
Happy and You Know It
Happy Birthday
Head and Shoulders
Hey Diddle Diddle
Hickory Dickory Dock
How Much Is That Doggie
Humpty Dumpty
Hush Little Baby
I'm a Little Tea Pot
Itsy Bitsy Spider
Jack and Jill
Let's All Sing Like the Birdies Sing
London Bridge
Miss Molly
Mulberry Bush

Favorite Children's Songs

Old King Cole
Old McDonald
On the Good Ship Lollipop
One Two Buckle My Shoe
Open Shut Them
Pat a Cake, Pat a Cake
Polly Put the Kettle On
Ring Around the Rosey
Rock A Bye Baby
Row, Row, Row Your Boat
Scales Do, Re, Mi, Fa, Sol
Skip to My Lou
Slippery Fish
Teddy Bear Teddy Bear
Ten Little Indians
The Farmer in the Dell
There Was An Old Lady Who Swallowed A Fly
This Little Piggy
This Old Man
Three Blind Mice
Twinkle Twinkle Little Stars
Wheels on the Bus
When You Wish Upon A Star
Whistle While You Work
You Are My Sunshine
Zip-a-de-do-da

10

Birthdays, Holidays and Special Occasions

Holidays and other special occasions can be sad rather than happy when families and friends are separated. Travel during holidays is often expensive and unpleasant. Video calling will take you where you want to be. Using Skype or FaceTime video call can help you celebrate special occasions despite the distance.

With your computer, you can participate in birthday and holiday celebrations. You can share traditional food, activities, art, and music. Video calls can also help the child build tolerance for others when you show respect for and teach about other cultures and customs. Many resources are available on the Internet to help you find creative holiday ideas and activities.

Visit www.apples4theteacher.com, a fun and educational website for teachers and children where you can find holiday activities and crafts.

Books for children about the holidays will help you talk about holidays, and you can read them to a child online. Check your local bookstore, or search on the Internet. Type in **holiday books for children** or **children's books about . . .** and name the holiday. I found more than twenty-four thousand books!

Birthdays

Celebrate birthdays with a video call to sing ***Happy Birthday.*** Watch candles being blown out and gifts being opened. With portable computers, laptops and phones, you can join in by watching the fun of a children's birthday party.

Holiday Ideas

New Year's Day

The start of the New Year is often a time of hope, remembering the past, and planning for the future. January 1 is not the beginning of the New Year for all cultures, so adjust the date for your family. Children often enjoy celebrating the beginning of a new year with hats and noisemakers. Read *P. Bear's New Year's Party: A Counting Book* by Paul Owen Lewis (1999) to preschool children.

Dr. Martin Luther King Jr. Day

Discuss who Dr. King is and why his life is celebrated with a holiday. Read *Martin Luther King Jr. Day* by Trudi Strain Trueit (2006).

Valentine's Day

Make valentines and talk about love and your love for the child. Read *The Story of Valentine's Day* by Nancy Skarmeas (1999).

Presidents' birthdays

Discuss which presidents have special holidays and why we celebrate them. Wear red, white, and blue or have a flag displayed in view. Read *President's Day* by David Marx (2002).

St. Patrick's Day

You don't have to be Irish to talk about this holiday. Tell or read a story about St. Patrick, elves, fairies, or leprechauns. Talk about the shamrock and wear green. Read *Saint Patrick's Day* by Gail Gibbons (1995).

Skype and Facebook have teamed up to create holiday greetings that can be sent for birthdays and holidays. Launch the ***Say It with Skype*** app on Facebook to send a holiday greeting.

Holiday Ideas

Easter
Color Easter eggs while the child watches or watch the child color and decorate an egg. Show some eggs that you decorated before the call. Parents can hide a few eggs, and you can watch the child hunt for them. Talk about the religious meaning of Easter and why we celebrate the coming of spring. Read *What Is Easter* by Michelle Medlock Adams (2006).

Memorial Day
Discuss the meaning of this holiday; when did it start? Read *Memorial Day Surprise* by Theresa Golding (2004) or *The Wall* by Eve Bunting (1992).

> DISCUSSIONS WILL VARY GREATLY DEPENDING ON THE AGE OF THE CHILD.

Fourth of July
Talk about celebrating the country's birthday and play or listen to patriotic music. You can display a flag in the room where you are sitting. Read *History for Kids: The Fourth of July for Kids* by Ian Fraser (2013).

Ramadan
During the ninth month of the Islamic calendar, you might discuss the practice of fasting, talk about the Eid al-Fitr festival, and read *Night of the Moon: A Muslim Holiday Story* by Hena Kahn (2008)

Halloween
It is great fun for children to be dressed in a costume but even more fun for them to show their costume to family. Some children use Halloween as a time to discuss their fears.
Read *The I'm Not Scared Book* by Todd Parr (2011).

Holiday Ideas

Veterans Day

Discuss war and peace. Talk about the importance of honoring those who serve to keep us safe from harm. If parents are on active duty, they may be able to discuss their work. Grandparents who were in the military services can share some of their experiences and pictures and show uniforms. Read *Veterans: Heroes in Our Neighborhood* by Valerie Pfundstein (2012) or *The Wall* by Eve Bunting (1992).

Thanksgiving

Discuss thankfulness and share projects that may have been made at day care and in school. Read *The Thankful Book* by Todd Parr (2012) or *Happy Thanksgiving, Curious George* by H. A. Rey (2010).

Hanukkah

Participate in part of the Menorah tradition, count to eight, watch the child spin the dreidel, and sing songs. Read *Celebrate: A Book of Jewish Holidays* by Judith Gross (2005).

Holiday Ideas

Christmas

Help decorate the Christmas tree by sending some of your own family ornaments or gift ornaments to share. Watch as the tree is decorated. Sing or play carols with the children. Read *Bear Stays Up for Christmas* by Karina Wilson and Jane Chapman (2008) or *The Christmas Star* by Marcus Pfister (2004)

Kwanzaa

Discuss the meaning of the word Kwanzaa and the African culture. This is a great opportunity for children to learn about a holiday that celebrates family, friends, and community. Read *My First Kwanza* by Karen Katz (2003).

Scholastic has published a series of Rookie Read-About Holiday books, with titles for more than fifteen different holidays; some have been listed previously.

IF YOU HAVE A FAVORITE WAY TO CELEBRATE A HOLIDAY OR BIRTHDAY USING VIDEO CALLING, SEND IT TO ME AT WWW.VIDEOCALLWITHKIDS.COM.

Conclusion

Your Journey to Discovering Video Call

If you haven't experienced video calling with children, try it today–you will love it. Don't delay and miss memories that will be in your heart forever. If you live miles away or are a parent who is traveling, the Internet makes it possible for you to see those you love no matter where you are.

Download a video call application like Skype using the easy steps in this book, or have someone do it for you so that all you will have to do is select the Skype icon and sign on. If your family and friends are Apple computer users, try FaceTime. Enjoy spontaneous visits while you sit back, relax, listen, and observe. Try the ideas in this book to plan fun activities that will work with the age of the child whom you are calling.

Savor the memories of seeing children grow, gifts opened, books read, and songs sung. These precious moments are the joy that video calling can give you.

For more ideas visit my website www.videocallwithkids.com and my blog www.videocallwithkids.com/connect

Technical Terms You May Want To Know

Technical Terms You May Want To Know

Technical Terms You May Want to Know

Don't run—you don't need to memorize the following terms; they will simply help you understand the basics of computing and video calling, and before long, they will be a part of your regular vocabulary. Refer to this glossary if you come across a term that you don't already know.

APPLICATION
An application, or "app," is a computer software program that helps you perform a task.

BROADBAND
The term broadband is short for broadband Internet access and refers to a high-speed Internet connection. Cable TV is an example of a broadband connection.

CAPTCHA
An acronym for Completely Automated Public Turing test to tell Computers and Humans Apart, used in computing to determine whether or not the user is a human.

DOWNLOAD
To download something from the web means to copy it from a website to your computer. Many things can be downloaded: data, books, movies and software applications (also called programs).

HEADSET
A headset is a device that is worn on your head and leaves your hands free. Headsets can come with a speaker for one ear, similar to a telephone, and are also available with speakers for both ears. There are headsets that need to be plugged into the computer and wireless headsets that don't need a connection.

ICON
An icon is a symbol that represents the application (or part of the application) that you want to use.

INTERNET

The Internet is an enormous networking infrastructure. It connects millions of computers together globally, so any computer can communicate with any other computer as long as they are both connected to the Internet.

INTERNET BROWSER

A web browser is a program that allows you to go to websites. Internet Explorer, Netscape Communicator, Apple Safari, Opera, Chrome and Mozilla Firefox are all web browsers. You will have one or more on any computer or smartphone that you purchase, and can download any others you choose for free.

MENU

Sometimes called a drop-down list or drop-down menu, this menu is a list of options or choices. Menus are accessed by placing your mouse or finger on the menu item.

SEARCH ENGINE

Search engines are programs that search for specified keywords and return a list of the places where the keywords were found. A search engine is a program found in all Internet browsers, that enables users to find information on the World Wide Web.

USB PORT

Short for Universal Serial Bus, this is where you connect an electronic device, such as a keyboard, mouse, webcam, or headset, to your computer. It supplies the power to the device.

VIDEO

The term video relates to visual images; originally pertaining to the transmission or reception of television images, later video technology included video tape and video clips for computers.

VIDEO CALL
Video call encompasses face-to-face visual and audio communication.

VIRTUAL VISITATION
Virtual visitation, sometimes called electronic communication or e-communication, can include video conferencing or video call, email, and instant messaging, as well as communication posted on the internet.

WEBCAM
The term webcam is a combination of two words, web representing the World Wide Web and cam for video camera. A webcam is a camera that is installed on a computer, tablet, or phone or an external camera that can be attached to the computer monitor or sit on a desk.

WEBSITE
A website is a page or group of interconnected pages that are found on the Internet.

WI-FI
This is a popular technology that allows for the transfer of information wirelessly using radio waves. Wi-Fi is a trademark of the Wi-Fi Alliance and is used as part of brand names for products.

WORLD WIDE WEB
The World Wide Web or simply Web, is a way of accessing information over the Internet. The web utilizes browsers to access information on web pages as well as to access graphics, sound, and videos.

YOUTUBE
YouTube is a website on which videos can be shared with others.

Internet Resources

www.apples4theteacher.com

Chock full of kids' activities, this teacher-created site provides information about parenting, education, early childhood, and home-schooling activities

www.babycenter.com

This site offers helpful information about pregnancy and parenting.

www.bootsonground.com

This website is dedicated to helping support the troops and their famlies during their deployment. Information about video call can be found on this site.

www.cdc.gov/ncbddd/child

The US Government's Center for Desease Control and Prevention provides research-based information about child development.

www.fbi.gov/fun-games/kids/kids-safety

This page of the U.S. Federal Bureau of Investigation (FBI) website provides safety information for children when they are using the computer.

www.fbi.gov/scams-safety/fraud/Internet_fraud

Another page from the FBI website, this one provides Internet safety information for adults and seniors.

www.internetvisitation.org

Provides updated information for separated and divorced parents about Electronic Communication/Virtual Visitation legislation. Includes tips for working with an attorney to establish virtual visitation.

Internet Resources

www.parenting.com

This is an informative site sponsored by the Meredith Corporation publisher of Parenting, Family Circle, and other magazines.

www.pbs.org/parents

This is an outstanding resource that provides information on child development and early learning and also gives ideas for educational games and activities.

www.raisingchildren.net.au

The Australian Parenting Network is an excellent resource about parenting and child health information covering ages zero to fifteen.

www.skype.com

The official website for Skype is the place to go for downloading the application that allows you to video call. There is also good support and help information on the site.

www.skypito.com

A free download of Skypito is available at this site. Skypito is the first safe online video chat communication tool for young users aged two to fourteen. It was designed to simplify and enhance the computer experience for children.

www.supportapple.com

Apple's official site for product information, support and downloads.

www.youtube.com

YouTube offers videos that can be posted by companies and individuals from all over the world. Find how-to videos about downloading and using Skype for video calling on this site.

Helpful Reading

Abdulezer, Loren, Susan Abdulezer, and Howard Dammond. *Skype for Dummies.* Indianapolis, Ind.: John Wiley, 2007.

This book offers advice about hardware, directions for downloading and installing Skype, and ideas for using the many Skype features.

Allyn, Pam. *What to Read When*. New York: Penguin, 2009.

A great resource for books and stories to read with your child at different ages and stages of development.

Bowers, Ellen. *Everything Guide to Raising a Toddler.* Avon, U.K.: Adams Media, 2011.

Written by the owner of a preschool, covers all the challenges of the toddler years, including the use of technology and computers.

Coleman, Paul. *How to Say It to Your Kids.* Paramus, N.J.: Prentice Hall, 2000.

Filled with specific advice on how to talk with your children about difficult topics, such as divorce, drugs, sex, and other complicated issues.

Einon, Dorothy. *How Children Learn Through Play*. Hauppauge, N.Y.: Barron's, 2004.

This child development expert provides a wealth of information about age-appropriate activities for two-to six-year-olds.

Einon, Dorothy. *Play with a Purpose.* London: Duncan Petersen, 1985.

Einon presents hundreds of suggestions for games and play that help stimulate learning for children aged six weeks to ten years.

Helpful Reading

Favaro, Peter. *Smart Parenting During and After Divorce.* New York, McGraw Hill, 2009.

An essential resource to help divorcing parents talk to their children about divorce and to deal with conflict-related to issues of custody, visitation and finances.

Gough, Michael. *The Virtual Visitation Handbook.* 2013.

A helpful resource for divorcing parents with specific information about legislation for virtual visitation/electronic communication. PDF format can be downloaded at www.internetvisitation.com.

Leach, Penelope. *The Essential First Year.* New York: Dorling Kindersley, 2010.

This book offers a comprehensive look at the first year by an international expert on child development.

Mayes, Linda, and Donald Cohen. *The Yale Child Study Center Guide to Understanding Your Child.* New Haven, Conn.: Yale University Press, 2002.

A comprehensive yet practical guide to child development from the prestigious Yale Child Study Center, founded in 1911.

Murkoff, Heidi, and Sharon Mazel. *What to Expect—The Second Year.* New York: Workman, 2011.

This source contains good descriptions of the milestones and behaviors of the stage between twelve and twenty-four months of age.

Sears, William, and Martha Sears. *The Baby Book.* New York: Little, Brown, 2003.

Written by a doctor and nurse, a husband-and-wife team, this is a comprehensive book on baby care, development, and behavior.

Helpful Reading

Shelov, Steven, ed. *The American Academy of Pediatrics Caring for Your Baby and Young Child—Birth to Age 5.* New York: Bantam Books, 2009.

In addition to guidelines and milestones for physical, emotional, social, and cognitive growth, this extensive book answers questions about child care and many other parenting concerns and issues.

Stamm, Jill. *Bright from the Start.* New York: Gotham Books, 2008.

A science-based way to encourage brain development from the director of New Directions Institute for Infant Brain Development.

About the Author

Dr. Lillian Tibbles is an author, teacher, psychiatric/mental health nurse practitioner and family therapist. She received both a master's degree in nursing and a doctoral degree from the University of Connecticut. She then completed postgraduate studies in psychiatry/mental health at the University of South Florida. She is the author of several articles about the use of computers for patient education.

During the tenure of her career, computers that were once found only in business environments, have now become commonplace in nearly every household. Recognizing this outstanding resource in a more personal way, Dr. Tibbles, through Skype and video call, began to communicate with, and immensely enjoy her grown children and young grandchildren who are hundreds of miles away.

Realizing that for some who are not confident using computers, attempting to video call might seem overwhelming, Dr. Tibbles decided to write this book. Her hope is that others will, as she has, enjoy sharing important life experiences with their loved ones.

For more information, please see her website at www.VideoCallWithKids.com

Index

Index

Index

Notes

Notes

Made in United States
Troutdale, OR
11/03/2026

56844712R00066